Conscious Robots

Conscious Robots

If We Really Had Free Will, What Would We Do All Day?

Paul Kwatz

Peacock's Tail Publishing

ISBN 978-0995069703

Amazon.com Reviews

"An absolutely necessary book."

"Should be taught in schools."

"Dynamite, this is a brilliant book."

"Bad ass!"

"Easy to understand and persuasive."

"Fun, short, insightful."

"It's an easy read, and lays out the bare truth with good humor. Let's get that 'Happiness Party' started ASAP!"

"This book presents a unique, concise argument for the validity of determinism from an evolutionary standpoint. Bravo to the author, whoever he may be!"

"Paul has an idea that he wants to share and you'll do yourself a disservice if you don't consider it."

"If a thoughtless ramble were a tweet..."

"The most useless book I have ever looked at."

If you enjoy reading *Conscious Robots,* you can review it simply by going to bit.ly/ConsciousRobotsReview.

Publisher's Note, April 2017

In March 2017, Jeff Bezos, soon to be the World's Richest Man, announced he'd be selling $1 billion Amazon shares every year to invest in his Blue Origin space rocket company.

"Each to his own," you might think. "Whatever makes you happy."

But having read *Conscious Robots,* you might think "Well, that's not going to work. I can see how it could maximise the survival chances of his genes but I can't see how it will fool his subconscious into making him any happier."

Conscious Robots changes everything.

And it changes nothing because nothing can be changed.

You might enjoy it because it will take you to the logical extreme. As one reviewer commented, the book "throws out the baby, bath water, tub and the sink."

Or you might enjoy it because it clears up a lot of the things in life that previously didn't make any sense at all.

But then you might hate it because, despite the promise, *Facing up to the Reality of Being Human* is pretty difficult and frustrating (but you'll blame your own brain for those unpleasant feelings, of course).

The best I can promise is that in 107 minutes time, you might think you have a better way to spend $75 billion than Jeff Bezos.

Introduction

3000 years ago, the Earth was flat and the stars were gods.

The gods ruled the Earth, and they created humans as their playthings, rewarding the lucky ones with rich harvests and victorious battles, and punishing the rest of us with floods, earthquakes and nasty diseases.

If you caught one of those nasty diseases, there were no reliable cures - incantations and prayers would be required to drive out the evil spirits, and if that failed you'd have to sacrifice a goat. Or maybe even a child.

Fortunately, we've come a long way since then.

The Earth is no longer flat, and the gods have delegated the majority of their powers to "natural phenomena". We still sacrifice animals to cure our diseases, but those sacrifices take place in the laboratories of science rather than on the cold stone altars of Thor.

We've even worked out why we exist. We can trace the origins of the Universe back to the Big Bang, and we can see how the ensuing cascade of physical interactions could place the stars in the cosmos and the Earth in its orbit around the sun. We can see how primitive life forms could emerge from the chemical swamp on the Earth's surface, and gradually evolve to become creatures so complex and sophisticated that they can worry about the size of their own backsides.

We get it. We understand the world, we know what's going on.

And yet…

And yet…

We're still living on a Flat Earth. Or, rather, its modern equivalent. One day, we'll look back and say "Did they really think like that in the 21st Century?"

It's not that we haven't done the science yet. The science is clear. What we haven't done is **accept what the science is telling us**. It's as if we've gathered all the evidence that the Earth is round... but we can't quite come to terms with such a crazy idea. "Don't be daft. You can see for yourself - it's obviously flat," we mutter, looking away and changing the subject.

So what is this modern equivalent of a Flat Earth?

Well...

There's no easy way to say this.

We're robots.

Not robots made of metal.

We're robots in the sense of being **automatic machines**, doing nothing more than **following instructions**... from the day we're born until the day we die. No freedom. Just doing what we're programmed to do.

It's not the easiest idea to live with.

So one of the aims of this book is to help you "live with it".

To help you accept that you are no more responsible for your actions and "choices" than the sun is responsible for the time it rises, a volcano is responsible for erupting, or the clouds responsible for the rain.

The book will also try to convince you that the thing you should be **worrying about**... (if you could actually *choose* what to worry about)...

... is the **mechanism by which your choices are controlled**,

... the mechanism that your programmers use to **fool you into working for their benefit and not your own.**

10

It's really the most elegant of con tricks.

We're programmed by millions of years of evolution to believe that **The Unpleasant Things in Life** - the things we work so hard to avoid - are caused either by

-        **uncontrollable events in the outside world,**

-        or **our own failures;** our own failures to "conquer" and get control of what happens to us in that outside world.

We're programmed by evolution to believe that the reason life isn't an endless stream of **joy and contentment** is because **we haven't *won* yet**:

- *that if we had more money, our worries would slip away*
- *that a fine house and elegant possessions would satisfy our desires*
- *that if we had a successful career - and children! - we'd be content.*

Ha!

And we fall for all of it (especially the children thing).

In short, we're conned into believing in "happily ever after".

But "happily ever after" won't ever happen.

And not because we can't achieve all the things we set out to achieve.

"Happily ever after" won't happen *regardless* of what we achieve - because it's simply **not in the interests of our genes.**

It's not in the interest of our genes to *allow* us to be satisfied with what we've got... no matter how wealthy, successful or "lucky in love" we are.

Instead, **our brains have been wired up to make us feel permanently dissatisfied.**

Even if our lives look wonderful to almost everyone else, we will always want more. It's why millionaires strive to become billionaires, it's why pop stars take drugs, it's why perfect couples end up divorced. And it's why humans have conquered the planet.

Happiness isn't the reward for a life well-lived...

...it's the mechanism used to control our behaviour.

This is the reality of being human.

Businessman Becomes First to Take "Happy Pill"

Yesterday, Jake McClure, Chairman of Genasoft and one of the world's richest men, announced that he would be "retiring from ordinary life" and consigning himself to the so-called 'Happiness Bed'. Although others are believed to have experienced 'The Bed of Roses' for short periods, McClure is thought to be the first to commit for the rest of his life, and is certainly the first to talk openly to the world's press. McClure, along with 'a dozen or so of the world's wealthiest individuals', is said to have been a heavy investor in the research for the last 5 years.

McClure, 43, briefly answered questions from the press.

Could you tell us about the research project that has led to this possibility?

Well, that's quite confidential. I've been a contributor for the last five years, but the project has been operational in various forms for longer. Many top scientists are involved, including several Nobels - initially it was hard to convince them of the validity of the project, but that's changed recently.

Aren't you sad to be leaving your wife and children?

Yes, indeed. I am very sad. But I know that the feeling of sadness will not be with me tomorrow. Tomorrow, and for the rest of my life, I'll know only joy, delight and profound contentment.

Isn't your life pretty good anyway?

Yes it is. But it's nowhere near as perfect as you might imagine. We all assume that huge wealth, great success and a wonderful family will elevate our lives to a degree of satisfaction and delight not experienced by so-called 'normal' people. We assume such a life will be free of fear, free of pain, free of worry. But it isn't. My brain is programmed always to want more than I already have: I cannot free myself from the desire for more, nor from a feeling that the grass could be even greener. And when I take a break from wanting more, and I count my blessings and reflect on how wonderful my life already is... I worry about losing all the things that make it wonderful.

Besides, even if my experience of life were significantly more wonderful than yours, you have to understand that the intensity of the feelings that I'll be experiencing tomorrow will be way beyond what is normally available in a sustained fashion to a human being - joyous feelings that in our normal lives are rarely glimpsed will become my minute-by-minute reality. I can't wait.

Won't you be bored?

No. Boredom is a device used by the subconscious mind to influence the choices made by the conscious mind. The Bed rewires the brain. I won't be bored, simply because my boredom circuits will be disabled.

Don't good feelings in life come from overcoming difficulties?

Indeed they do. That's absolutely the way it works in 'normal' life: our brains only give us satisfaction when we overcome difficulties. But these good feelings that we crave are nothing more than neurological events - patterns of neurons firing in particular ways. Currently, our brains are wired up by millions of years of natural selection so that we experience these feelings only when we improve our genetic survival chances. But that's just wiring. Our brains can be rewired, just like a light bulb can be rewired. We've simply put in a new switch... only this time it's going to be permanently in the 'on' position.

Aren't you concerned about the impact on society?

Certainly when the technology becomes more affordable there will be profound changes to society in general, profound changes to people's priorities. The impact on the economy will be interesting to see... but I won't be around to see it, because it won't be as interesting as actually being on The Bed. Am I concerned about that? As long as no-one unplugs me, I'll be fine!

Aren't you being a bit selfish?

Absolutely. But I've discovered that my fear of being regarded as selfish isn't sufficiently powerful to forego a life of barely-conceivable bliss.

Are you indeed the first?

I'm the first to publicise, yes.

But not the first?

At this final question, Mr McClure smiled and turned away from the reporters.

Quick Poll Question

Would you want to spend the rest of your life in blissful happiness, without ever having to move again?

Definitely / Maybe / Absolutely not!

Vote at <u>consciousrobots.com/16</u>. **Find out if you're alone in wanting to plug yourself into a happiness bed…**

Part 1:

How the Illusion Works

*Why we **believe** we're making our own choices... when in reality we're just doing exactly what we're told to do.*

Part 1: Robot

.

Happiness is the meaning and the purpose of
life,
the whole aim and end of human existence.
- Aristotle

Part 1: Robot

The Fundamental Mystery of Being Human

Usually science solves mysteries. But when it comes to the question of how humans acquired their "free will", our expanding knowledge of the world seems to have created a mystery bigger than any that it's solved.

For thousands of years, philosophers and scientists have debated whether humans can really have the freedom to choose what to do with their lives. For a discussion to remain entertaining and unresolved for such a length of time requires both sides to be highly persuasive, and indeed they are – the two arguments are so convincing that it's hard to imagine a weakness in either of them. Yet one of them must be wrong, because they are direct opposites, incompatible.

- On one side is our *personal experience* of being human.

- On the other side is… "science".

Our personal experience tells us that we are "free to choose" – free to choose our religion, our politics, whether or not to commit a crime.

Whereas *s*cience tells us that we are "bags of chemicals" - our bodies and brains composed entirely of atoms… and nothing else. Inside each of us, these atoms react together to make the body function; from the firing of nerve cells in the brain to the contraction of the muscles in the legs, it all comes down to a series of highly complex - and highly ordered - chemical reactions.

> *If you were to pick yourself apart with tweezers, one atom at a time, you would produce a mound of fine atomic dust, none of which had ever been alive but all of which had once been you.*
> *- Bill Bryson*

There is, of course, no "freedom of choice" in a chemical reaction: when two atoms react together they do it automatically. They don't think about it, or wait for instructions from a brain, they just do it.

According to our understanding of the physical world, our whole lives can be nothing more than an "inevitable cascade" of chemical reactions, beginning at the moment that our father's DNA met our mother's. Or, to go back even further, beginning billions of years ago with the Big Bang, at which point all the movements of all the atoms in the universe were set in motion.

> *The initial configuration of the universe may have been chosen by God, or it may itself have been determined by the laws of science. In either case, it would seem that everything in the universe would then be determined by evolution according to the laws of science, so it is difficult to see how we can be masters of our fate.*
> *- Stephen Hawking*

No one has a problem with this concept when it's used to predict the precise time of the next solar eclipse, or to calculate the age of the universe. No one would want to get into a jumbo jet whose atoms didn't behave in the entirely predictable ways that those atoms behaved when Boeing were doing their testing...

But what about when you apply this basic chemistry to the life of a human being?

The conclusion is absurd. Impossibly bizarre. Of course humans are much more than the chemical reactions that make them up. Of course we're free to make up our own minds. It's quite obvious that we are **genuinely able to decide what to do next.**

But... *how?*

If we really do have the freedom that we're so convinced we have, then the human brain would be the only known structure that has the ability to gain control of the chemical reactions that make it up.

So the Free Will Challenge is "How do we reconcile our personal experience of 'being alive' with all our accumulated scientific knowledge?"

The way we resolve this dilemma is to trust our personal experiences. Because our perception of being able to choose is **so powerful**, we assume that it must be the science that's wrong: "We know we're free," we argue, "so there must be a scientific explanation that we've yet to discover."

But then again, our instincts have been wrong before. We used to think that the world was flat and the stars were gods.

So rather than overturn the laws of physics, wouldn't it be simpler to *challenge our experience of free will?*

What if our ability to choose is an **illusion**? If it could be shown that when we think we're making a choice, we're actually *following instructions...* wouldn't we have solved the age-old mystery and reconciled our personal experience with decades of scientific progress?

How the illusion works

We're about to take a closer look at what happens when we make a decision. Not at the molecular interactions that cause neurons to fire and electrons to whizz round our brains, but rather at the thoughts and feelings that lead us to make one particular choice over another.

By examining the processes we go through when making such a decision - whether it's to change the TV channel or to jump off a cliff - it becomes surprisingly obvious that all we're doing is following instructions.

Then, by examining *where these instructions come from*, and how they got to be programmed into our brains in the first place, we can make a direct link to the cascade of atomic interactions that began with the Big Bang.

And we can begin to face up to the idea that we're robots.

Part 1: Robot

Chapter 1: How We Make a Decision

To analyse whether or not our perception of free will is an illusion, we need to take a closer look at the thought processes we go through when we're making a supposedly free-willed choice.

The science writer, Matt Ridley, has summed up our instinctive feeling that we're able to choose what we do with our lives:

> *I am quite capable of jumping in my car and driving to Edinburgh right now and for no other reason than that I want to.... I am a free agent, equipped with free will.*
> - *Matt Ridley*, 'Genome'

And this appears to satisfy Dr Ridley. He knows he's got free will because he can drive to Edinburgh simply because he wants to...

But how does he know whether he "wants to" go to Edinburgh or not?

We can imagine his thoughts as he mulls over whether to make the trip:

> Perhaps he's heard what a beautiful place Edinburgh is and that the shopping's rather good. Maybe he has a friend there, or possibly he just wants to prove that he's got free will. These would be the *incentives* for the trip - the benefits to be gained by going to Edinburgh.

But pretty much everything we want in life has some sort of cost:

> It's not clear how far it is to Edinburgh, but we'll assume it's a fair distance. So Dr Ridley will have to take into account the price of petrol, the boredom of the long drive and if he's anything over 6ft, the backache from being wedged into the inadequate space behind the steering wheel. These are all *costs* of the journey - things arguing against him taking the trip.

But how does he know whether the downsides of the trip outweigh the advantages? If it were a simple financial transaction, he could simply compare the monetary cost against the profit. But how do you compare the "cost" of a backache with the "profit" of seeing a beautiful city?

The answer is that you have to think about how it will make you *feel*.

We humans are equipped with the ability to *imagine in advance* how much pleasure or pain a particular event will give us: it's what we do when we're umming and ahhing about a decision – we're trying to imagine what are the most likely outcomes, and to find out how good or bad those outcomes might make us feel.

So when someone is "deciding for no reason" whether to go to Edinburgh or not, what he's actually doing is weighing up:

> - how many **"good feelings"** he's going to get from the trip: how much **satisfaction** picking up a bargain in the shops, how much **pleasure** from sight-seeing, how much **contentment** at proving that he's got free will.

Against:

> - how many **'bad feelings'** it will cost him to get there: **aching** back, **frustration** at slow traffic, **anxiety** about the cost of petrol, **concern** for the environment.

If the scales come down on the side of **good** feelings, then he's in the car and on his way. Thinking about how he "feels" about potential outcomes tells him which of the potential outcomes he actually "wants" to achieve.

Quick Poll Question

Imagine you are Dr Ridley: do you agree that simply being able to drive to Edinburgh means we have free will?

Strongly agree / Agree / Disagree / Strongly Disagree

Vote at <u>consciousrobots.com/26</u>. How do your thoughts compare to everyone else's?

Feelings are the currency of decision-making.

It's how we make any decision: we respond to how we *feel* about it.

There are two types of "feeling": feelings that we *like,* and feelings that we *dislike*.

- **Feelings we dislike**: guilt, sadness, despair, hunger, unhappiness, fear, anxiety.

- **Feelings we like**: satisfaction, pleasure, contentment, happiness, joy, sense of achievement.

When we change the channel on the TV it's because we're bored with the program currently on the screen. "Being bored" isn't a feeling that we put up with for long if we can get rid of it simply by pressing a button. When we donate money to charity, we do it because we feel "concerned". Feeling concerned, or "feeling sorry for someone" is a mildly unpleasant or disconcerting sensation that can be alleviated by handing over some cash.

Without an emotion, we simply wouldn't know which decision to make. Indeed, if we didn't have any feeling about a situation either way, we wouldn't *care* which decision we made... in which case we probably wouldn't have been thinking about it in the first place.

Part 1: Robot

.

**When we make a decision,
we're attempting to**

- *maximize* **our pleasant feelings**

and

- *minimize* **our unpleasant feelings.**

Part 1: Robot

This observation – that our choices are determined by our feelings – can be used to resolve the contradiction of whether or not we're really "doing what we want with our lives".

Firstly, consider the perception of freedom: if we chose to do what our feelings are telling us to do.... then our freedom all comes down to whether we are able to choose which feelings we experience in given situations. If we can't, then whatever it is that's choosing our feelings... is really making all our decisions for us.

Secondly, if it can be shown that the "thing" that's choosing our feelings is the mechanical, automatic process of chemical reactions that has been rolling along since the dawn of time, then our perception of choice is compatible with our being "just a bag of chemicals". We no longer have to assume that there's something special going on inside our heads that gives us our unique freedom.

In the following chapters, we'll look at *why feelings exist*, what their purpose is and what they're designed to make us do. In doing so, we'll show that human choice could indeed be nothing more than a pre-determined series of chemical reactions.

But first... given that we're disposing of our free will on the basis of this single observation, how can we be so sure that our choices are controlled by our feelings?

Do feelings really control our choices?

> *A man can surely **do** what he wills,*
> *but cannot **determine** what he wills.*
> *- Schopenhauer*

The Evil Scientist

While Dr Ridley is deliberating whether or not to go to Edinburgh, let's imagine that he somehow falls into the hands of.... *an Evil Scientist.*

The Evil Scientist gains access to Dr Ridley's brain and finds a way to **rewire** his neural pathways so that **she** – the Evil Scientist – is in control of whether he feels good or bad about something. She could, for example, set up his brain so that he was *terrified of the thought of going to Edinburgh...* so that everything that was previously attractive about his proposed journey now filled him with dread and fear.

Would he still go? He'd be somewhat confused that the thought of the visit was now making him feel bad, but would he put himself through the effort of the trip if he could see so much pain coming from it?

He would of course still *feel like he was making his own decision....* but in reality, he'd be doing precisely what the Evil Scientist told him to do.

But surely we're capable of deciding how we feel about something for ourselves?

We might consider that we have the ability to "talk ourselves" into feeling good: that mature, well-rounded individuals are actually in charge of their own emotions… but it's not an easy task.

The only power we have is **in response to a feeling we already have.** We can't simply *choose* never to be sad again – we can only attempt to control a sadness that we're already experiencing.

> From the moment we wake in the morning, we're being assailed by feelings we'd rather not be having – the insistent ring of the alarm clock reminding us of the painful reality that we've got to get up and face the day. We can't stop the bad feelings arriving, otherwise we'd just tell ourselves to be delighted at the start of each day: all we can ever do is try to "think positive", and maybe distract ourselves by turning the radio on.

If we really *were* capable of controlling how we feel about something, then wouldn't we all be a lot happier than we already are? If we could control our own feelings, then why would any of us ever allow ourselves to experience misery and disappointment again?

*But I often choose to do things I **don't** want to do - like working for an exam, or putting in overtime…*

We often do things we don't want to do. Indeed, some people claim they spend most of their *lives* doing things they don't want to do. But is that really accurate? Are we really choosing options that will make us feel bad when we *could* choose to do something that would make us feel good?

Let's examine *why* we do things we don't want to do.

It could be that *neither* of the options open to us will give us any pleasure: nobody likes paying taxes, but we usually hand over the money because the penalty for *not* paying them would be even more painful. What can we do other than choose the thing that hurts *least*? It's the best way to minimise the total unpleasant feelings we're going to experience.

Or maybe we've calculated that the way to get the best result out of any particular situation is by doing things that hurt a little bit first. We're "investing in the future" - working hard now because the ultimate satisfaction will exceed the effort put in. Study for your exams because you'll be better off with the qualification; go jogging because it feels better to be slim and fit. Psychologists call it "delaying your gratification" and it's one of the things that children have to learn as they grow up. Babies don't do it, adults do. 10-year-old children are somewhere in between, on the painful path of learning that often the way to get the *most* out of life is by doing things that hurt a bit first.

*I often do things I don't want to do in order **to help other people** - simply because it's **the right thing to do**.*

We talk about doing things "because it's the right thing to do" - and by that we usually mean we're doing something we wouldn't choose to do, other than out of a sense of duty or moral responsibility. The implication is that we more than simple pleasure-seekers, we're able to overcome our baser instincts.

But aren't we *still* choosing the option that will ultimately lead to the least pain?

> My friend calls and says, "I really need your help to get a piano up the stairs to my flat." I weigh up the options. I'm going to miss the pleasure of my favourite TV programme, there'll be the effort of driving over to my friend's house, I imagine slipping on the stairs and being hurled down three flights with a piano on my head. I'm formulating an excuse in my mind... when I start to feel *guilty*. I imagine my poor friend struggling to get the piano up the stairs on his own. I think of all the favours he's done for me in the past, and I realize I can't win: helping makes me feel bad, but **failing** to help makes me feel worse. I feel so bad at the thought of my poor friend struggling without me that before I know it I'm pulling on my shoes and rushing out the door. I've chosen the action that will ultimately give me least pain. It's what makes me such a nice person.

When we care about other people and things like the environment, the point is not so much *what* we care about, but rather that we *care at all*: although one individual might care about money and fame whilst another individual cares "unselfishly" only for others, the point is that both are "caring" – both are experiencing a *feeling*. And unless they're able to *choose for themselves* precisely what feelings they experience... then their choices are being *controlled by* those feelings.

Somehow it seems deeply insulting to say that we help other people simply to make ourselves feel better.

Did Mother Theresa *dislike* helping the poor of Calcutta? Presumably what made Mother Theresa a "saint" was the realisation that - for her at least - the greatest pleasure, satisfaction and contentment would come from working for the benefit of others. The effort and the difficulties would be far outweighed by the rewards. Do we have a problem that it made *her* feel good to make *other people* feel good?

To be motivated to help another person, we have to know that they have problems – that they themselves are in pain. But we can only care about their pain by feeling pain ourselves.

Quick Poll Question

 Do you agree that we help other people only because it is the option which makes us feel the least pain in the long run?

Strongly agree / Agree / Disagree / Strongly Disagree

Vote at <u>consciousrobots.com/35</u>**. Are you as pragmatic and calculating as everyone else, or do you stand alone believing that we help people just to be nice?**

Summary

Regardless of whether our behaviour is generous or selfish, it would seem that the action we choose depends our inner feelings.

As Mark Twain put it:

> *When there are two desires in a man's heart he has no choice between the two but must obey the strongest, there being no such thing as free will in the composition of any human being that ever lived.*
> *- Mark Twain, "Mark Twain in Eruption"*

Chapter 2: The Purpose of Feelings

We've observed that whenever we make a decision, we're choosing to do things that we think will make us feel good. But that wouldn't mean we're robots, would it? After all, if we've got the ability to choose to do things that we think will make us feel good... then that's what being free is all about, isn't it - doing what you want with your life?

To understand whether we are robots or not, we need to investigate *why feelings exist* - what their purpose is and *whether they work to our advantage*.

Why does food make us feel good? Why does being loved make us feel good? Why does being admired make us feel good?

We've all lived with feelings from the day we were born, they're as a much a part of our lives as our arms and legs. But whereas it's easy to see what arms and legs are for, we don't often ask ourselves "What's the *purpose* of my anger?" or "I'm happy today - but what's the *function* of this happiness that I'm feeling?" Our assumption is that feelings don't have a *purpose* at all – that they are *results*. But perhaps we're too busy *reacting* to our feelings, rather than *analysing* them.

Where are feelings created?

One of the first things we might notice when we start analysing our feelings is that they aren't actually being *created* by the situations that we experience... they're being created by our own brains.

Two brothers are watching a football match when the team in blue scores a goal. One of the men leaps around enjoying highly pleasurable feelings of delight, whilst his unfortunate brother experiences profound anguish and distress. The same event in the "outside world" has created two entirely different experiences in the "inner world" of feelings.

Although "happy" and "sad" seem to be **automatic responses** to events in the world around us, these feelings aren't somehow being beamed into our brains from the outside world, like an alien beaming instructions into our heads with a ray-gun. When we climb a mountain and admire the view from the summit, although it seems to be the *view itself* that's making us feel good, that's not really how it works; the view has no capacity to control how we feel because the view is nothing more than a collection of different wavelengths of light passing through the atmosphere and entering our eyes... It's up to our brains to *interpret* that information and convert it into our experience of "enjoying the view".

A feeling is not an inevitable consequence of the physical world - a feeling is created inside our own brains.

Indeed, there is nothing inherently pleasurable or painful about *any* event in the world that we see, it's all just light waves, all just electromagnetic radiation. Our brain has to **decide** whether an experience should be pleasant or unpleasant.

Which must mean that:

Our own brains decide whether *anything* we experience should make us feel good or bad.

The astonishing reality is that our own brains must be deciding how much we enjoy our *whole lives*:

Our own brains decide whether our *whole lives* are utter delight or abject misery.

It doesn't seem right somehow. Our own brains are *choosing* to make us feel miserable?

> *When a prisoner is trapped in his enemy's torture chamber and subjected to hours of horrendous pain, the pain he experiences is **created entirely by his own brain**. Granted, he wouldn't be experiencing that horror without the physical abuse, but the torturer would have no weapon were it not for the body's **decision** to create these enormously awful feelings.*
>
> *We blame our torturer ... but should we be blaming our own brain?*

It seems bizarre to suggest that our own mind somehow *chooses* to make us feel pain when we're being tortured. But there can't be much doubt that somehow and for some reason, the brain is wired up so that torture is experienced consciously as "horrific pain". But why should this be so?

Indeed, how does our brain know when we should feel good, and when we should feel bad?

Hold that question for a moment.

What do we mean by "me"?

We've been trying to understand whether or not "we" are free to make our own choices. But who, precisely, is this "me" that we are all so aware of?

When we think of who "we" are, it feels like "we" is our **whole body**... from the top of our heads to the tip of our toes, it's all part of the idea we have of "me". But it's more complicated than that: I am still "me" if I lose a leg, or if I lose an arm. I might have a different outlook on life as a result, but my basic personality remains in my brain, it doesn't leave with my departing arm. I am still "me" if I am confined to a wheelchair without the use of my arms or legs.

So it appears that "me" is not my really my *body* – but rather my *brain*.

And we can take it further. Because "me" isn't **all of** my brain... it's only **a part of** my brain.

I'm not really "there" if I'm **asleep**. My *brain* is there, but I'm not there myself - at least I'm not *aware* of being there myself. One part of my brain continues to work, keeping my heart beating, my lungs breathing, my temperature carefully regulated... but it's not something I'm in any way *conscious* of while I'm asleep.

My only experience of life - indeed my only experience of anything at all - is when I'm awake, when I'm *conscious*. Therefore, there must effectively be two parts to my brain: the conscious part and a part that *isn't* conscious. I'm not aware of any of the processes that my non-conscious mind goes through to regulate my heart rate or the temperature of my blood, therefore is it strictly "me"? Would it not be more accurate to say that "I" am not my *brain...*? "I" am only the *conscious* part of my brain? My conscious mind, thoughts, sensations and feelings are the true extent of "me".

We are our conscious minds.

Only. Nothing else.

The religious concept of "the soul" is a tidy comparison: the thought that, once the body dies, the soul lives on, rising above the body, floating around, somehow surviving outside the body, is so powerful and easy to understand because **consciousness is our only experience**. For "soul", read "conscious mind".

Which means that, when we consider *where* exactly our feelings are being created: in our brains, yes... but in our *conscious* mind... or in the part of our brain that is not conscious? Clearly, we *experience* these feelings consciously... otherwise we wouldn't be aware of them. But we don't *create* them consciously otherwise we'd just be able to feel great all the time regardless of our external circumstances. Therefore, although our feelings are *experienced* consciously, **they must be created non-consciously** – that is, in a part of our brain that is beyond our conscious control.

Feelings are created in a part of the brain
that is
beyond our conscious control.

Which has some considerable implications…

If…

the aim of our lives is
to maximise pleasant feelings and
minimise unpleasant feelings,

and…

whether we feel good or bad is under the control of a
part of our own brain that we can't consciously
control,

then…

**the aim of our lives amounts to nothing more than
convincing our own brain to make us feel good.**

Back to the question:

How does our brain know when to make us feel good, and when to make us feel bad?

We can now rephrase the question to be more specific:

*How does our **non-conscious** brain know when to make us feel good, and when to make us feel bad?*

To understand this, we need to go back to the origin of feelings, and the most basic feelings of all:

Pain

When you put your finger in a flame, it hurts. And it hurts for an obvious reason: your finger is getting damaged. Your body (or rather your non-conscious mind) is telling you to take action quickly if you want to keep your finger. Pain hurts.... but we couldn't live without it.

...and Pleasure

Although it might not be immediately obvious why we get pleasure from such things as the opera and Picasso, the pleasure we get from eating ice cream and donuts is much more easily explained: the fat, sugary food helps us survive. If our non-conscious minds weren't telling us which food is tasty, how would we know what to eat?

Somewhere way back in our deep murky past, our ancestors didn't have any feelings at all. They didn't experience pleasure, they didn't even know *what pain felt like*. They just... **did**. Somewhere near the very beginning, the earliest forms of life were single cells - like bacteria or amoeba. No nervous system, no pain, no pleasure.

And then feelings *evolved...*

As animals started to move around, they needed a quick method of knowing whether something was good or bad for them. The animals that felt pain when their bodies were damaged were the ones more likely to survive. Which meant that they were the ones that had more chance of reproducing and passing their genes on to their children. Animals that had **genes for pain** flourished, animals that **didn't experience pain** died out.

And the animals that experienced pleasure when they ate sugar and fat were the animals that had enough energy to find more food, to defend themselves and to stay warm at night.

> *[Brief aside – please don't read: this is an unsatisfyingly simplistic explanation. Organisms don't **need** to experience pain to react to something is bad for them. We can be reasonably confident that snails aren't conscious - and if they aren't conscious then they can't "experience" pain the same way that we do. But they are still capable of moving away from damaging chemicals: "moving away" is an automatic "chemical" response by the organism to the chemicals on the ground (but no more automatic than our conscious experience of pain and our "automatic" choice of what actions to take in response to that pain. It's still "just chemical reactions"). Experiencing pain came with the evolution of consciousness. So why did consciousness evolve? Probably because it allows us to imagine "what would happen if...?" And to decide whether "what would happen if" is something that will be good for the survival chances of our genes. And to be able to do that, we need to be able to imagine the pain or pleasure of something that isn't actually happening. Which might be why we don't see more snails travelling to Edinburgh.]*

Evolution appears to train us in the same way we might train a dog: we get **"little dog biscuits of pleasure"** every time we do something that's **good for our chances of survival**, and **"little chastisements of pain"** whenever something happens that's **bad for our chances of survival**.

Quick Poll Question

 Where do you currently draw the line between something being described as a "machine"?

At humans, because we are machines / At animals solely guided by survival instincts, such as lions / At unconscious animals such as snails / Single celled amoeba / Nothing living is a machine

Vote at <u>consciousrobots.com/43</u>. Find out how your view compares.

If something happens that is *good* for our survival chances - we feel *good*.

If something happens that is *bad* for our survival chances - we feel *bad*.

It's how we know what we need to do to survive.

Everything we feel? *Everything* we do?

*Free will is a delusion caused by our inability to understand
our true motives.*
- attributed to Darwin

Would it make sense that our evolutionary heritage is still somehow
controlling *everything* we feel... and therefore everything we do?

Well, yes. Indeed, it would be much more surprising if evolution *wasn't*
controlling all our behaviour.

Evolution couldn't simply **create us**, and then with a cheery wave of its
hand say "OK, *Homo sapiens*, that's *my* job done. It's all yours now.... feel
free to do what you want with your lives." Evolution couldn't have worked
like that.

The Washing Machine acquires Free Will

*Imagine you came home one night to find that your washing machine
had unplugged itself, wriggled out from underneath the kitchen
worktop and caught the bus into town for a beer with its mates.*

*You'd not only be somewhat surprised that your docile domestic
servant could grow arms and legs... but you'd also want to know
how it had developed a mind of its own and a thirst for a beer.
Washing machines simply don't decide to do things that they're not
programmed to do: they don't have the ability to become anything
more than their manufacturers made them to be.*

So what makes us think that we humans have this ability?

At **some stage**, if we go back far enough, there is no doubt that our
ancestors were machines. It's up to you how far you want to go - you
might be quite happy to believe that your ape-like ancestors didn't have
free will, or you might want to go back as far as single-celled amoeba-like
creatures, but go back far enough and we're all going to agree on the
"machine" diagnosis. Which means that at some point - and we don't really

care when - *something had to happen that gave our ancestors this ability to choose what to do with their lives.*

So the question is - what was the mechanism of change? *How* did we go from **being machines** to **not being machines**?

Surprisingly, there's no easy way to answer this. It appears to be impossible.

Why is it impossible for free will to have evolved?

Evolution is a two-stage process:

1) Firstly, there are **"random" changes,** either chance mutations in the DNA that create (unplanned) changes in the genes, or combinations of parental genes causing new characteristics or behaviours in the child.

2) Secondly, there is **selection**. Selection is the point at which all these random changes get **judged** to see if they're good or bad. Most changes reduce an individual's chances of passing on their genes to the next generation. Some changes make no difference at all, and occasionally some changes to the DNA turn out to cause **improvements in the functioning of the individual**; like being able to see further, to run faster or to fight diseases more effectively. Thus, these improvements **increase the individual's chances of survival** or of having children and thus increase the chances of these new genes being passed on to the next generation.

Now, even if we assume that it's actually possible for "free will" to have appeared by random chance (and we can only assume that for a brief moment), for a Free Will Gene to then have been **selected...** doesn't make any sense at all: "free will" is just about the last thing that would ever have been chosen by natural selection.

The world is a ruthless place. A baby hominid born with the free will to do "whatever it likes" isn't going to last very long out there under the gaze of hungry predators. The baby hominid that survived to become our ancestor was the one that didn't waste energy "doing its own thing" but was entirely focussed on the job of passing its genes on to the next generation. "Free will" would have been an insurmountable burden in the same way that "no eyes" would have been an insurmountable burden.

But an equally large problem is...

How do you create free will - is that even possible?

*How do you create a thing that can do what it wants to do, **without telling it what it wants?***

Let's say you manage to create a robot that is physically capable of wandering around the world looking for food. And let's say you want to give it the freedom to choose its food for itself. Just like you would with a human child.

You can't say to your robot "Get out there and taste things - see if you like them". It would just reply, "How will I know if I like them or not?"

You'd have to install some **criteria** that told the robot what was "nice" and what wasn't. You'd have to say something like "go find more food that tastes like broccoli" - just like you might do with a human child.

But if you did that, you wouldn't be giving it freedom to choose for itself, you'd be telling it to find food that tasted like broccoli.

To really give it personal freedom, you would need it to be able to *work out for itself* what's a nice-tasting substance and what's not.

But how could you achieve that? Or how could the robot achieve that for itself? You could program it with the skills to learn from its environment what was nutritious – to learn how to get the information it needed to make that decision: to read some books, to talk to some scientists, to create some sort of testing mechanism. But once again, you've still given it the criteria by which to make its decisions: you've told it to look for "nutritious" food.

Which is of course what evolution does to the human child. The human child is born *preprogrammed* to like high-calorie food and to dislike things without digestible calories.

That's not to say you can't also **educate** the child. You can convince it that broccoli might well be better for it long term than the ice cream that tastes perfect to it naturally. But the broccoli still has to be fundamentally nutritious. Try convincing a child that cotton wool tastes good. Or gravel.

You could say to your robot, "Go find things that are blue," or "wet", or "cold". But they'd all still be **your** criteria. To give it "free will", you'd need to be able to say, "Go and do whatever **you** want to do, Robot: find out for yourself whatever it is that **you** like." But that wouldn't work because it can't create its own criteria for what is good and what is bad, for what it "likes" and it doesn't like. The closest you could get would be to give it some sort of **randomness device**. "Go and find things at random." But "random" isn't choice either. And "random" isn't what we humans do. We're not **random**. We're extremely specific about what we want.

Quick Poll Question

Can we choose what we want to want?

Yes / No / Unsure

Vote at <u>consciousrobots.com/48</u>. Do you believe that, if we really wanted to, we could eventually choose to like to eat gravel? Or is that just a bit far-fetched?

Chapter 3: Slaves to Our Genes

OK - but even if our choices are controlled by our feelings, and our feelings are controlled by the need to enhance our survival chances, that wouldn't mean that we're robots, would it? Evolution's helping us survive - isn't that what we'd choose to do anyway?

Would evolution make us do things that aren't for our own benefit?

Pain and hunger are good news for us. They keep us alive. We don't mind experiencing a little bit of pain if survival is the reward.

But evolution isn't just about *survival*... take sex, for example:

At first sight, evolution's definitely looking like our best friend where sex is concerned: "having sex" appears without doubt to be a choice we'd make with our free will. Not only is sex a great deal of fun, but we also need sex to *survive*. We need sex to have children, and if we didn't have children, the human race would die out....

But hold on a minute... what does "having children" have to do with *us*? Us as **individuals**, that is? Sure, as a *species*, we do of course need children. And if every one of our ancestors hadn't had sex at least once, we ourselves wouldn't be here...

*...but what's in it for **us**? What do **we** get out of sex* - apart from the obvious pleasure, that is?

The truth is that each of us would probably *live just as long* if we never had sex again. We'd actually live longer, because sex is not only unnecessary for our individual survival, it's also *dangerous*. Especially for an animal in the wild.

The dangers of having sex and having children:

> Pregnancy is just about the most expensive and difficult job that any human gets. Not only is there the requirement to find more food to eat, but the mother's got a bigger weight to carry around when she's hunting and being hunted. And then there's the risky drama of *giving birth* itself, followed by the costs of providing for the offspring, defending it and eventually teaching it the ways of the world.

> Even the males don't get off lightly, frequently being required to fight for territory, or - if you're a peacock - to carry a ridiculous tail around with you just for a *chance* at being the favoured choice. And then there's *disease*. For both genders, sex is a terribly easy way to spread disease - no condoms or antibiotics for our ancestors.

> Kids! Who needs them?

Well, the point is - *we don't.* So why do we do it? Why do we have sex... and why do we have children? Why do our brains sweet-talk us into doing something so dangerous.... if it isn't for our own benefit?

Because evolution is not about us. *Life* is not about us. The reason we exist... isn't for our own benefit.

Evolution is about our *genes*. We exist as machines to help spread our genes around the world. It seems crazy, it seems like it's the wrong way around, but we can't deny that it's our *genes* that survive and get passed on, not us.

Compare yourself to a recording of your favourite music.

> What's "the point"? The music... or the device (vinyl, tape, cd, smartphone) that the music is stored on? Does the **music** have an existence because of the **device**, or does the **device** exist because of the **music**?

> Clearly, it's all about the music. The music can be copied to another storage device, its essential structure and sound unchanged as each device becomes obsolete and disposed of.

> The storage devices exist because they're a good way of spreading music around the world. They aren't the purpose, they aren't the point... they're the mechanism. Equally, human beings aren't the

point, we're just the device by which the genes get to be spread round the world.

Our genes are the music. We're the discardable and vulnerable vinyl discs.

Which is why we humans aren't designed to behave in such a way that *we as individuals* will survive.... we're designed to behave in a way that will help *our genes* survive. Our behaviour is controlled in such a way as to ensure that we maximise the survival chances of our genes. It's why we get pleasure from sex, it's why we get pleasure when our children do well at school.

If something happens that's bad for the survival chances of *our genes* - we feel bad.

If something happens that's good for the survival chances of *our genes* - we feel *good.*

Of course, that's not to say **it's irrelevant whether we live or die.** *But we're not alive for the sake of living.* We're alive to create the next generation. "Being alive" is just a mechanism to increase the chances of our genes surviving.

That's the way evolution works. It's what evolution *is.* You can't have the process of evolution by natural selection if you don't have genetic information that gets passed on from one individual to the next. If our ancestors had behaved in such a way that they maximised their own *individual* survival chances over the survival chances of their genes *themselves* might have survived longer... but they wouldn't have passed those genes on: the individuals with genes for "look after yourself first and don't have children" didn't reproduce. It was only the individuals that took the risk of having sex that *could* have become our ancestors. Natural selection can **only** select behaviour that favours the genes ahead of the individual.

We only like having sex and we only like having children because evolution "wants" us to like it. We're not *choosing* to like it.

But... *I like kids - I **want** to have kids.*
Having kids is precisely what I would choose to do anyway, regardless of whether I have free will or not. And I love having sex as well. You can't seriously be trying to tell me that it wouldn't be my free will choice to have sex?

What if we could get all the pleasure we get from having children or sex... without actually having to do it? Sounds absurd, of course... but if we've agreed that the only thing we care about is *how we feel*, we have to consider the serious possibility that if we had the opportunity to experience all the nice feelings *without the accompanying risk...* we'd be very advised to take that opportunity.

Quick Poll Question

If you genuinely had free will – what would you do all day?

Feel good / Survive / Pass on my genes / Something else

Vote at consciousrobots.com/52. Unless you really can't choose anything, in which case there's no point in asking.

Trapped inside an evolutionary machine

We are survival machines - robot vehicles blindly programmed to preserve the selfish molecules known as genes. This is a truth which still fills me with astonishment. Though I have known it for years, I never seem to get fully used to it.
-   *Richard Dawkins,* "The Selfish Gene"

If "we" are our *conscious minds...* then maybe we're not so much Dawkins' "survival machines"...

... as ***trapped inside*** survival machines.

We can't escape. We can't stop the whole machine doing things for survival purposes. Indeed, we've got to be active helpers in this whole job of spreading our genes because the only thing we want to do is to feel good, and the only way we've found to get to **feel good** is to help to increase the survival chance of our genes.

- But *our* purpose is not to spread our genes...

- *Our* purpose is to make ourselves **feel good...** for as long as we can stay alive.

If our conscious minds were really in control, wouldn't we simply instruct our *non*-conscious minds to make us feel good all the time? Wouldn't we do enough to survive - to live a full-length life – and then spend the rest of the time simply *feeling good?* There would be no need to reproduce, no need to "achieve", no need for luxury or success - indeed no need to do anything at all. If we could control our own feelings, wouldn't we choose quiet, risk-free lives of extreme contentment, delight, joy, satisfaction and well-being? Think how good it would be for global warming.

If we genuinely had free will - what would we do all day?

Would we just choose to **feel good?**

All day?

Would we set up our lives so that *everything* we did made us feel fantastic, regardless of the reality of the world - no pain, no remorse, no self-doubt, no guilt, no jealousy, no anxiety?

After all, once we've made sure that we were going to **survive**... what's our motivation for doing anything else? Anything else we do is just a method of ensuring that we're going to feel as good as we possibly can in the future.

So why not cut out the middleman - and just feel good. Regardless of what happens in the world around us?

What next?

If the aim of our lives is to have a pleasant, rewarding and satisfying existence then we need to be sure that we have a reliable way of achieving this goal.

The method that most of us use is the one that comes instinctively to us – to try to get the most out of the world we live in. We assume that a successful career, a loving family and a reasonable amount of healthy leisure time will result in happiness and contentment for the rest of our lives.

But is this really what happens? Maybe not. Part 2 will suggest that our dreams of living "happily ever after" are simply not possible. Not because we're incapable of meeting The Handsome Prince but because, even if we did, our evolutionary programming would ensure that as soon as we got back from honeymoon we'd have found a hundred other things in our lives that needed improvement. Part 2 will suggest that, in the words of Buddha, *life is inherently unsatisfactory;* not because the world is too difficult to control but because **our own minds will never allow us to be satisfied**, regardless of what happens in the world we live in. So if we want to find a way to achieve the goals of our conscious minds, we're going to need a different approach...

And what about God?

If humans do have free will - if we're not robots - then an unworldly power appears to be the only way this could have happened. The enormous scientific knowledge that humans have built up - of physics, of chemistry; the science that cures our diseases and flies us to the moon - does not allow for freedom of choice.

Unless that freedom is God-given.

Review of Part 1

There are two "facts" that appear to be incompatible with free will:

1) our bodies and minds consist entirely of automatic chemical reactions;
2) we were created by the process of evolution.

Whilst these both appear incompatible with *free will*, they are perfectly compatible with *each other* - evolution forming the link between the first chemical reactions bubbling away on the surface of the Earth and our existence in the 21st Century, showing us how our bodies and actions can have been created in a world that is nothing more than atoms following pre-determined, unbreakable rules. "Life" itself need be nothing more than an automatic process from conception to ultimate decay, DNA molecules giving the blueprint for the great cascade of chemical reactions that we call our lives.

But if these two "facts" are true, then how do we explain our experience of free will? How is it possible that we can spend hours agonising over a decision... when the choice we will eventually make has been pre-determined since the dawn of time?

An answer becomes possible when we examine our *experience* of choosing. If 1) our choices are controlled by our "feelings"; and 2) our feelings are controlled by evolutionary pre-programming, then "what we want" is nothing more than "what we calculate will maximise the survival chances of our genes". If our "choices" are simply responses to evolutionary pre-programming, then our behaviour can, after all, be entirely "pre-determined".

Even though we may deliberate for hours, this is just time spent in calculation. The choice we finally make is no more special than a choice made by a chess computer.

Part 1: Robot

Think you could program a computer to have free will?

Not convinced by Part 1? Do you still think that the fact our bodies and minds consist entirely of automatic chemical reactions *and* the fact we were created by evolutionary processes are both compatible with *free will*? You could win £1,000. One could argue that all actions we take are to "maximise the survival chances of our genes", but for such an argument to even be worth having, one has to be able to explain how "free will" could have arisen in an organism's brain. We think it's reasonable (as part of the discussion around whether free will could exist in an evolved organism) to see whether we can imagine how free will could arise in a computer. Hence the competition, and your chance to win the money.

The entries will be judged by Professor Steve Jones, Fellow of the Royal Society, influential biologist, broadcaster and author of numerous popular books on evolution and the natural sciences. Read the full details at consciousrobots.com/win, or scan the QR code with your phone.

Part 2

If We Really Had a Free Choice, What Would We Do All Day?

*If we're slaves, working hard to achieve the **goals of our genetic masters,** will we ever be allowed to achieve our **personal goals**? If we're controlled by the feelings we experience, what does that say for our chances of living enjoyable and satisfying lives?*

Will our evolutionary programming allow us to achieve the happiness and contentment we're all striving for? Or will we have to find another method to get what we want?

Chapter 4: Is Happiness Achievable?

How happy does evolution want us to be?

Chapter 4 is an attempt to get a better understanding of **happiness**... *the* **nature** *of happiness. Not – "How do I get to be more successful?" or "What are tonight's winning lottery numbers?" but more like* "**How does happiness work?** *What's the* **purpose** *of happiness?"*

If we can get a better understanding of how happiness works from a **biological** *point of view, we might have a better chance of understanding whether we're going the right way about trying to get happier.*

Or, indeed, whether it's possible to get any happier....

We're brought up with the idea that happiness is **achievable**. We're taught that if we work hard at school, if we have a great family, if we earn money and the respect of the community... we'll be happy. It's what "happily ever after" is all about: every fairy story we heard as a child was based on this simple assumption. And yet, one of the world's most popular religions was formed 2,500 years ago by a man that wasn't convinced.

Buddha was an Indian prince. According to legend, he was a wealthy and important man... but it didn't make him content. In search of "more" out of his life, he sat down under a tree to try to work out what was wrong. After much sitting, he finally came to the conclusion:

"Life is inherently unsatisfactory."

Whatever you achieve, whatever you have in life – be it material possessions, love, family, respect of people around you – *it's never going to be enough to make you content.*

This is a bit shocking to the 21st Century Western mind. The thought that millions of people base their spiritual beliefs on the idea that *getting what you want* won't satisfy your desires seems somewhat bizarre.

Could it be true?

Born 2,500 years ago, Buddha had very little scientific knowledge to help him with his deliberations. Although he had an *observation*, he didn't have an *explanation*. So what can more than two millennia of science add to Buddha's observation?

Could it be **human nature** that we're never satisfied? Not in an old wives' tale kind of way, but in a *biological* way? Could it be "human nature" in the same way that it's a *lion's* nature to kill antelopes? If so, would that mean that there's a scientific explanation for why the grass is always greener? A scientific explanation for why life is *inherently* unsatisfactory?

How our instincts mislead us about happiness

We know that feelings aren't simply automatic results of what's going on in the world around us, they're actually *mechanisms to control our behaviour* - mechanisms used to make us do things that will increase the survival chances of our genes. So if *happiness* is a mechanism, rather than an automatic result... where does that leave our chances of living happily-ever-after?

We'll consider two Personal Experiences that show how our instincts mislead us about the way happiness "works". Then we'll see whether these observations, combined with what we know about evolution, can improve our understanding of happiness.

Quick Poll Question

Do you think life is inherently unsatisfactory, like Buddha?

Yes / No

Vote at <u>consciousrobots.com/64</u>. It probably won't make your life any less unsatisfactory, but at least it will be interesting to see everybody else's view.

Personal experience Number 1: *The Pleasure Fader*

The Man with one music album.

There's a man in Halifax, Nova Scotia, who owns just one record album. It's by Bruce Hornsby, but that's not important. What *is* important is that, unlike the rest of us, *this man doesn't **get bored** with his one album*. He just plays it over and over again.

His friends say to him "How can you possibly play that same record over and over again, Bill?" And Bill just smiles, and says, "How come you can't?"

OK - he's fictional. And he'd have to be, wouldn't he? After all, that's frankly inhuman. However good the music, however talented the performer, to be able to listen to just one recording, over and over for ever - it's not normal.

But *why* do we get bored? Why is it that we can play a piece of music five times while we're learning to love it, then maybe ten times more, enjoying it every time... before, ever so gradually, the music loses its charm?

- Nothing changes in the music: we don't *wear it out* by overplaying it, like we might wear out a pair of our favourite shoes. Nothing changes in the outside world: the sound waves don't hit our eardrums in a different way...

And yet, *we get bored*. For some reason, **our own brain decides to take pleasure away from us...** and make us go out and buy another album. Exactly like the music industry wants us to.

And, it all seems *so normal. "That's just the way it is. Some things will never change."*

Personal Experience Number 2: *The Expectation-Adjuster*

The Christmas Bonus

Two employees receive a $5,000 end-of-the-year bonus. One of them flips burgers for a living, and the other works on Wall Street. Mr Burger Flipper is *delighted* with his $5,000... whilst Mr Wall Street is mortified. Last year, his bonus was $200,000.

The men's reactions are entirely dependent on their *expectations of life*. They both experience exactly the same event... but one brain converts the information into *"delight"* while the other brain converts the information into *"misery"*. And it all seems so normal. *Of course* the Wall Street man should be upset with his $5,000 - he's worth much more, he's capable of much more...

But hold on - surely *success should make us happier?*

The Wall Street trader is **"successful"**. He's the one that people on a lower wage are supposed to aspire to be. And yet, he's the one that's *unhappy*. If happiness works the way we think it does, shouldn't we be happy if we do well - *regardless of what we already have?* If happiness is a reward for "success" in life, then "how good we feel" shouldn't depend on our *expectations*... it should depend on our *success*.

- Mr Wall Street should feel *equally* happy with the $5,000 bonus as Mr Burger Flipper does...

- And if he gets the $200,000 he was expecting, he can be *even happier*.

But of course, it doesn't work that way. If we think we're *capable* of a big bonus, we get upset if we don't get it. *And it all seems so normal.*

We know the Pleasure Fader and the Expectation Adjuster so well that we don't even question them. They are so much a part of our lives that we blame the world we live in, and not the world inside our heads. Yet without these devices, how wonderful - how perfect - our lives would be.

What's the purpose of the pleasure fader and the expectation adjuster?

Until we realise that we're genetic slaves, we don't even *look* for an explanation. We blame the world we live in. But the world we live in isn't responsible for how we feel; it's our evolutionary-programmed minds.

So, would it make sense that *evolution* should have created the pleasure fader and the expectation adjuster?

Imagine you're evolution...

Imagine you're evolution. You've been working hard for the last few billion years, and recently launched your greatest creation so far, *Mr Homo Sapiens*...

Things are going well. Your new machine is beginning to throw its weight around the planet to great effect: with his new **agriculture** project, food shortages are rarely a problem; with his new **tools,** he's successfully defending himself against all the nasty animals that are trying to eat him; with his new-found ability to **harness fire**, it's warm and cosy at night and he's learning how to cook.

Indeed, your new creation is becoming so successful that his circuits are overloading with feel-good chemicals. It's one big holiday for *Mr Sapiens*. He's working an hour or two a day, and the rest is... well, sitting around smiling.

He's gone from *wanting*... to *having*. And it's changed everything. The state of mind that made him get things done, take risks, push for results... has gone. He's lost his motivation.

Which means there are problems coming.

Surviving in **the natural world** is like surviving in **business**: however successful you might already be, there's always someone trying to steal your market. If you sit back, holding on to your *old methods*, your *old prices* and your *old technology*, pretty soon some young upstart is going to start offering a cheaper, faster, greener service.

In business, you don't have time to enjoy your past successes. And it's the same in the natural world. There's always some other animal trying to take your food, there's always a distant relative trying to steal your mate.

So, if you're evolution, and you're trying to program your machine to be as successful as possible, it looks like you need a rethink:

You motivated your new machine so that it would work hard in pursuit of happiness... but you didn't work out what was going to happen **when it got happy.**

How do you make sure that your favourite creation is **always pushing for more?** How do you make sure that happiness doesn't take away his desire to succeed, his desire to improve things? And how do you program the machine so that it will **always achieve the maximum that it's capable of achieving?**

Firstly, you've got to make the pleasure fade. And secondly, and just as importantly, you've got to make those little dog biscuits of pleasure *relative to previous achievements*: you can't be congratulating a child that's just crawled across the room... if it's already capable of walking.

*Your machine can't feel good whenever the situation is **"good"**...*

*... it can only feel good when the situation **gets better.***

The Expectation-Adjuster on the golf course...

What makes a sport like golf endlessly entertaining? Is it that there's always a chance to improve? A continual challenge: no matter how good you are, you can always get better. Which is, of course, where the satisfaction comes from - when you play better than you thought you were capable of.

And even though you keep thinking, *"If I could just sort out my putting... I'd have this game mastered..."* You've been thinking that for years. You've got a lot better, but there's always more better to get. And fantastic golfers are no happier with their game than good golfers: everyone gets to feel good *relative to how much they're improving,* not to how good they actually are. A game that's equally absorbing for the beginner and the expert.

…And equally *frustrating*:

If you don't play as **well** as you expected, you feel the **pain** of failure. And there are no freebies. Unless you're **upset when you lose**… you don't get pleasure when you win.

In golf, you're not playing your opponent, you're playing yourself… or are you playing your evolutionary-programmed Expectation-Adjuster?

And golf is like life…

Everything we do in life is subject to an adjustment of our expectations. From the bonus at work to how we feel when we look in the mirror, satisfaction and contentment are dependent on our expectations.

Such a simple solution, such a simple device to make sure a robot is always achieving the most that it's capable of.

How else could evolution have set us up?

- Evolution didn't know whether we were going to be born in a penthouse on Fifth Avenue, or in a slum in the Third World.

- Evolution didn't know whether we were going to find life a complete breeze… or whether it would be like walking into a Force 10 headwind.

So, it had to come up with a mechanism that kept us motivated *regardless of our situation.*

And what could it do, other than simply ignore our *actual* situation… and consider instead only whether or not our situation was *improving?*

If your situation in life is *improving*, evolution *rewards* you.

If your situation in life is *falling*, evolution *punishes* you.

Imagine you lived 100 years ago...

You get up in the morning... and it's cold. The central heating must be broken. But of course, there *isn't any central heating*. There's also no hot shower. Never mind, once you get into the car, you'll warm up and it'll be nice and cosy at work. But of course, there's no car, and you've got to walk to work, and it's no warmer at work because you work in a field. The only thing left to look forward to is getting home and putting your feet up in front of the TV...

Most of the things we take for granted in the 21st Century didn't even **exist** 100 years ago. Can you imagine a life without cars, TV, good hospitals, fridges, fast food...? It's almost impossible to believe that so many of the things that we rely on for our happiness today weren't available **at all** to our great-grandfathers.

But did that mean they were **miserable** all day? Did that mean their lives were barely liveable? Or were they just as happy as we are?

What have we gained with our relentless pursuit of improvement? We live longer, for sure. But are we *happier?* Do we **need** this expectation-adjuster that seems to affect everything we do, everything we dream of? Wouldn't we be so much better off *without it?*

Unfortunately, the expectation-adjuster has never been **programmed out of us,** so we keep on building faster cars, taller skyscrapers, smarter mobile phones... all in an endless attempt to get more satisfaction than we got when we were cavemen.

Summary

What's the best way to get happy?

For most of us, the answers are obvious, instinctive. We don't need to think about it, we just *know* what will make us happy. The problem is not with the *deciding*, it's with the *doing*.

But if we're robots, programmed by natural selection to maximise the survival chances of our genes, then maybe "happiness" isn't the automatic result we've always assumed. Personal experiences seem to indicate a problem: "feeling good" has much less to do with our *actual* situation in life, and much more to do with whether our situation is *improving*. Pleasures fade and expectations adjust, leaving us always wanting more than we have.

Even before we apply the theory of evolution to the problem, there seems to be little doubt that however much control we get over the world we live in, we're never going to experience more than occasional sensations of bliss. Profound satisfaction is never going to be our constant waking experience.

How robot are you?

Fancy knowing whether you're certainly a robot, programmed by evolution - or perhaps lean more towards being human...? Find out by taking the quiz at bit.ly/How-Robot, or by scanning the QR code below with your phone.

Part 3:

The (Inevitable) Future

Can we ever overcome our robotic programming and take control of our lives?

*Can we achieve conscious control over **everything** we feel?*

*Would we **want** to?*

Chapter 5: Humans Fight Back

*Our instinctive approach to achieving our aims guarantees a life that is inherently unsatisfactory, programmed always to want something that we haven't got. It's not surprising that we humans have been trying to get **direct control** over our feelings for thousands of years.*

From daily attempts to "look on the bright side", to disciplined meditation techniques, we're never far from a conscious attempt to improve how we feel.

But perhaps it's **drugs** that have been our most effective weapon: from the mild results of alcohol and paracetamol to the more extreme effects of cocaine and heroin.

So what can the most powerful of these drugs tell us about our chances of overcoming our evolutionary programming?

Heroin

Take the best orgasm you've ever had, multiply it by a thousand and you're still nowhere near it.
*- Irvine Welsh, "*Trainspotting*"*

Heroin has a very similar chemical structure to substances found naturally in our bodies called *endorphins*.

Exactly what endorphins do is poorly understood, but they seem to have a role in the "reward circuits" in the brain. They are known to be released during exertion and stress, possibly as a mechanism to dull pain to allow us to maintain high levels of effort: if you're being chased by a pack of lions, there's no evolutionary advantage to being hampered by the pain of a stitch. Endorphins are also thought to be responsible for the feelings of euphoria often referred to as the "runner's high" experienced during intense exercise.

Most of our knowledge of the action of endorphins comes from their structural similarity to heroin and related molecules such as morphine. When heroin is injected into the blood stream, the brain can't distinguish the heroin molecules from endorphin molecules, resulting in intense pleasure, euphoria and elation.

Unfortunately, our evolutionary programming doesn't let us enjoy this new pleasure for long. Already present in the brain is a system for mopping up the naturally produced endorphins (to make sure our pleasure fades so that we can get back to working for the genes). This mechanism quickly absorbs the heroin molecules, the euphoria fading almost as quickly as it fades in "real life".

Not easily deterred, the conscious mind, continuing its bid for freedom, attempts to restore the pleasure by injecting more heroin. And this works for a while. But not for long. The brain makes an adjustment, adapting to the artificially increased quantities by *changing the ability of the body to respond to endorphins* – and also therefore to injected heroin. The body becomes "tolerant" of the increased levels... and life is inherently unsatisfactory.

For the heroin taker, this tolerance is good news only for his dealer: more and more heroin needs to be injected to achieve the same effect. But the worst problem is not that the user must take increasing amounts of heroin, it's what happens a few hours after each "hit". The body has become tolerant not only to *heroin...* but also to the *naturally occurring endorphins.* All the minor aches and pains that are normally dulled by endorphins are now free to persecute the conscious mind unchecked. The user begins to "withdraw" and starts to feel lousy. He's become tolerant of his own endorphins, and there's no heroin left to counter the effect.

A vicious cycle begins: the more heroin you take, the more it hurts when you're *not* taking it. And pretty soon you're "addicted" - you feel bad under normal circumstances and you remember how fantastic you felt when you were taking heroin. If the addict can't obtain more heroin, or has sufficient incentive to stop taking it, he experiences the "cold turkey" of withdrawal.

Thankfully, these cold-turkey symptoms don't last. Once again, the body is able to adjust - this time the expectation adjuster works the other way and the endorphin levels produced by the body return to their original levels. The tolerance is removed and the brain restores its normal response to normal levels of endorphin. Life is inherently satisfactory.

Heroin – friend or foe?

Heroin is almost universally disapproved of: if it could be eradicated like smallpox, most people would consider that to be a great benefit to the human race.

And yet...

Could it not be said that heroin is perhaps the conscious mind's best attempt at overcoming the slavery imposed by our genes? When used to control the excruciating pain of surgery or in terminal cancer patients, opiates such as heroin are without doubt some of our greatest friends; only when they get in the way of our "normal life" do they become dangerous pariahs.

But what are we trying to achieve in these "normal lives" we cling to? We cherish our "normal lives" because we believe they'll make us happy. But what if we could live just as long lives... but instead of experiencing the fear, disappointment, sadness and boredom of "normal life" we could experience the highs of heroin... all the time?

Part 3: Future Happy

.

Chapter 6: Would You Take a Happy Pill?

The Future

Our hero - disheveled but ruggedly handsome - emerges from the sewers of New York where he's spent 20 years living with mutant crocodiles.

He discovers a world of horror - machines have taken over the world. The entire population of humans are lying in coffin-shaped metal pods, electrodes protruding from their heads, wired up to silent computers. At first sight, and to the un-heroic eye, they appear to be enjoying their incarceration; the unearthly noise hanging over the city is the sound of a million people quietly chuckling to themselves. As our dismayed hero moves swiftly from one tortured individual to the next, the chuckles escalate into uncontrolled giggles, culminating in bloodcurdling gales of laughter.

Our hero - his rugged brow creased with the burden of being humanity's last hope - struggles to turn off the machines and free his fellow citizens. Tearfully, he greets his weakened comrades as they struggle from their cubicles and seize him passionately. But to his surprise, our conquering hero isn't born aloft in gratitude. Instead, humanity seems to be more than a little upset with him. They seize him roughly, shave his head and pin him down inside his own little box. He struggles, he fights… but he is but one and they are many. As the electrodes bite into his skull, he grits his teeth and steels his mind to repel the horrific brainwashing he's about to experience.

And it's working! His mental strength is too much for this simple machine. A triumph! The thrill of success courses through his body, a thrill like he's never felt before. He's at one with himself, a Renaissance hero after all, he can do anything – anything he damn well likes. He is all-powerful. And he chuckles with the joy of it all. As the chuckles escalate into delighted giggles, he realises he's free! Free from the dark forces! Free from the tyranny! Finally content, finally at the end of his journey, he barks with laughter as he discovers what it is to be truly alive.

His fellow citizens shake their heads in affectionate amusement. "Well, most of us had to try it before we believed it…" they sigh as they hurry to their pods, quickly plugging electrodes back into their heads and cranking the power up to 11.

Taking Control

Is it possible that humans will one day conclude that the only thing that matters to them are the feelings experienced by their conscious minds? Not because they are selfish or badly brought up, but rather because they are built that way, their conscious minds programmed by millions of years of evolution to pursue feelings of pleasure and satisfaction.

Will they at the same time become painfully aware of the inherent weakness in the traditional method by which their conscious minds attempt to achieve this aim – realising that no matter how much control they achieve over the world they live in they'll be no closer to achieving satisfaction and contentment than their caveman ancestors?

And will their minds turn to a new focus, towards a technology that allows them to control the world *inside* their heads rather than the world outside?

The Perfect Life?

Ever dreamed of the perfect life?

> *You've got homes around the world, a yacht in the Caribbean and more sexy sports cars than you can count. You've just invented a cure for cancer, your friends are film stars and you travel the world promoting charities for sick children...*

Not everyone's idea of a great way to live, but we all have our dreams. And the more optimistic we are, the more outlandish our aims, the less chance we'll ever get to *live* them.

But what if... instead of actually *living the dream* - we could just *enjoy how it felt?*

> - You don't actually *own the yacht*... but you feel just as good as if you did. The President isn't *really* on the phone for advice on foreign policy... but you feel the same sense of satisfaction as if he were.

After all, who needs the actual experience? Surely all we need are the feelings that go with it...?

> *"And now, ladies and gentlemen, thanks to the miracles of modern science, you too can experience the perfect life...*
>
> *"Just climb up onto this hospital bed, and we'll put a small drip in your arm and these rather comfortable electrodes into your head.*
>
> *"No, don't worry, madam, you won't need a book to read – you can just lie there, feeling better than you've ever felt before - no worries, no fears, no guilt, no pain, just perfect, unremitting **bliss**.... for the rest of your life."*

Would you do it?

It's rare to find someone whose immediate reaction is "Yes, of course I would. What else is life about?"

Most people think they'll somehow get *bored*, or that "you can't get happy unless you deserve it". They think that taking a pill to get happy is a cop-out or a weakness that would inevitably lead to profound dissatisfaction. Very few people would be willing to give up all the things that are currently important to them: their children, their gardens, their sense of personal identity - in return for what appears to be "meaningless" easy pleasure.

Everyone's got an opinion about that hospital bed, and the answer's usually "No, thanks."

A Brave New World

In 1930, Aldous Huxley wrote a book that has come to symbolise our instinctive reaction to the proposition of a happy pill:

> In *A Brave New World*, Huxley describes a future in which a pill called *soma* is used to control the population, to make them content with their lot. People take the drug from the day they're born until the day they die - they're not on hospital beds, they continue with normal lives, but *soma* relieves them of the burden when it all gets a little too *difficult*. Unfortunately, this relief comes at a cost. Not only does it remove the lows of life, it removes the highs as well – there are none of the thrills, none of the elation that we enjoy so much. Life's never difficult... but it's never a joy either.
>
> Only the handsome hero suspects that there's anything more to life. He's the only one with the courage to face the difficulties of life without the drug, demanding the right to be unhappy.

Huxley presents the story as a warning against "designer drugs" and their potential use by an oppressive government. In doing so, he accurately mirrors most people's instinctive fear of a happy pill.

But let's be realistic. Huxley's pill wasn't much to get excited about. It was more of a tranquilliser than a happy pill, and it came at a high cost. *Soma* was modelled on some of the drugs that were available on prescription in the 1920s. It was susceptible to *tolerance*, and overdoses were possible. If soma was the best drug that Huxley could come up with, no wonder he didn't want to take it.

Perhaps if he had suggested a drug that was **significantly better than life itself**, and *then* given us good reasons why it would be inadvisable to take it, *A Brave New World* might work as a coherent argument against a "happy pill". But as presented, it's nothing more than a warning against indiscriminate use of drugs that don't work very well.

The *real* happy pill, when it comes, won't be a *mild sedative*. It won't be something that pacifies us when we're bored…

The *real* happy pill, when it comes, will feel like…

…Armstrong felt when he took his One Giant Leap

…Mandela felt when he took his long walk to freedom after 17 years on Robben Island

…Churchill felt when Hitler's body was found in Berlin.

This drug will activate the parts of our brains that normally only get moving when we win Olympic gold, when we get elected President, when we find out that our cancer is cured.

After all, it's all just neurons firing.

Quick Poll Question

Would you want to feel happy, even if you didn't "deserve" it?

Yes / No

Vote at <u>consciousrobots.com/83</u>. Maybe by voting, you'll feel happy *and* deserve it…

> # It's all just neurons firing.
>
> There's nothing else going on in that brain of ours.
>
> - When we fall in love... it's just neurons firing.
>
> - When we experience the *best moment of our life* (and let's hope it hasn't happened yet)... it's just neurons firing.
>
> - When we're so upset we could just *kill* someone... it's just neurons firing.
>
> And even when we're so upset we could just kill *ourselves*... it's *still* just neurons firing.

So why do we readily and easily reject the idea of a pill that would give us a better life than the world allows us to experience?

It's too easy. Where's the challenge in that? It reminds me of the time when I was ten and my dad let me win at tennis. There was no sense of satisfaction.

Pleasures and satisfaction in life come when we achieve something, when we're sufficiently involved in life, sufficiently dedicated to have made something happen that's actually worth feeling good about. There's "no gain without pain".

We know this because it's our personal experience of life. In our experience, "feeling good" is only associated with achievement, with overcoming struggles. But the reason it's associated with achievement is because that's how we're wired up... because we're evolutionary robots, and that's what it takes to improve our genetic chances.

But the happy pill - or more likely the hospital bed and the electrodes - will not be like "real life". It will be a rewiring of the computer, cutting out the whole mechanism that at the moment is controlling the release of the pleasure, and - more importantly - cutting out the expectation-adjuster.

I want to feel sad when it's the right time to feel sad

If your child dies and you don't feel sad, it seems to imply that you didn't love them, that you're a bad parent and a bad person. But would your dead child have *wanted you to feel sad?*

Evolution wires us to feel sad when a child dies so that the next time we're in a situation where someone that we love (someone evolution has *told us* to love) might be in danger, we'll take greater steps than we might have done before.

Evolution decides when the time is right for you to feel sad, not you.

Part 3: Future Happy

A row of flashing buttons

Still not convinced that you want to be happy? That there isn't something *more to life*?

> *"OK... We'll make your hospital bed a little more **sophisticated**, then. You get **a row of buttons**. Each button representing a feeling...*

> *"There'll be buttons for satisfaction, buttons for achievement, a button for how it feels when you jump out of a plane and another button for how you feel when your parachute fails... Hey, you could even have a panel of celebrity chefs cooking up perfect recipes of emotions to delight your palate.*

> *"And whenever you want to experience a particular feeling, all you have to do is hit the corresponding button: if you want to feel like you've just overcome a particularly thorny problem at work, then we'll have a button that makes you feel precisely like that. When you want to know what it feels like to win a gold medal, you can experience it...*

> *"And if you really want to feel sad... it's this little button right here."*

Hard to believe that sadness button is going to get a ***great deal*** of use.

Part 3: Future Happy

Heaven on Earth

The happy pill was around a long time before Aldous Huxley.

Some religions call it "heaven".

When we die, our body decays, and our soul floats upwards to spend eternity in bliss. No pain, no suffering, just great feelings.

Heaven isn't a concept most of us have a problem with. Few say, "Don't send me to heaven - it all sounds a bit too easy." Few say, "Aren't we going to be a bit bored in this heaven place?"

For "soul", read "conscious mind".

For "heaven", read "happiness bed".

Part 3: Future Happy

The Happiest Man in the World

The happiest man in the world is a prisoner. He's been locked up in a hole in the ground for the last ten years. He gets barely enough food and water to survive, and only 30 minutes of natural light every day. His body is gradually withering away, despite his efforts to exercise. He is the happiest man in the world.

This man experiences almost complete joy every day. He was lucky enough to be born with a brain disorder. No matter what happens to him, he just feels great. It doesn't seem to matter what his torturers try, he just loves it all.

Possible. Just about possible that such a man could have been created by chance - in one of the random variations thrown up by evolution. He would never be *selected for:* he just wouldn't function as well as "normal" men who suffer the desires and pains of their evolutionarily-functional brains.

As an evolutionary machine, he'd be one of the *least* successful. But as an individual conscious mind, he'd be the *most* successful... because his conscious mind would have achieved its programmed aim - of being happy.

He'd be the most successful conscious mind that ever enjoyed the earth.

Part 3: Future Happy

Happiness Party Sweeps to Power

As polls closed last night, it became clear that an unprecedented change had occurred in British politics. Ever since Oliver Cromwell began the slow route to democracy, political parties have promised to improve the economy and the wealth of the people - the assumption being that an improvement in the standard of living would automatically lead to an improvement in the quality of life experienced by the citizens. Voters have now emphatically rejected this assumption. From now on, it appears, governments are expected to make their constituents happy. *Directly.*

It's been a slow route to power for the Happiness Party. 20 years ago, when Jenny Fastnet first began her campaign she was regarded as a harmless nut. With grace and humility, Fastnet gently insisted that our internal feelings are what's important, and that "how we feel" is rarely linked to economic prosperity.

Her claim that humans need to be saved from the manipulative powers of their own brains fell on closed minds for over a decade until drug companies and neuroscientists began to publish research that suggested that chemical control of conscious feelings was moving from rumour to reality and the voters began to want a slice of the fun for themselves…

Quick Poll Question

Would you vote for the Happiness Party?

Yes / No

Vote at <u>consciousrobots.com/93</u>. Is your political stance shared by the majority?

Final Word

"We" - our conscious minds - are trapped inside an evolutionary machine.

Regardless of our achievements in the world, we are condemned to our unfair share of misery, disappointment, heartache and despair. We're condemned to miss out on the sustained heights of delight and joy that our minds are capable of experiencing.

But one day we'll discover how to control our feelings, and then we'll enjoy a quality of life never experienced by any human before us.

We must not *fear* it:
    - to fear it is to misunderstand our existence.

And we must not *legislate* against it:
    - to legislate against it would be to subjugate millions to unnecessary pain and suffering.

And when the great day comes, we'll still be robots.

But we'll be doing exactly what we'd do all day if we weren't.

But what about the aliens?

Why haven't aliens visited the earth yet?

There are 70 sextillion stars in the observable universe.

22 zeros.

It's been 14 billion years since the Big Bang.

Only 9 zeros.

But plenty of time for the intelligent life-forms that **"surely must"** have evolved on those uncountable solar systems to develop a way to "reach out" across the expanses of time and let us know that they exist.

It's a paradox. Fermi's Paradox, because Enrico Fermi was the first intelligent life-form (on this planet) to realize that these enormous numbers didn't tally with the complete lack of Alien in our lives.

There are more than 20 theories as to why Aliens haven't contacted us yet (only one zero, but a laudable number of explanations nevertheless):

> – *they lack advanced technology, periodic extinction by natural events, it's the nature of intelligent life to destroy itself ...*
> *...they are here undetected...*

But here's the real reason.

Before all those uncountable Aliens worked out the physics of how to do the "reaching out", they worked out that there was no point. Because it wouldn't make them any happier.

They are out there.

There are millions of planets full of millions of lifeforms far more intelligent than we are.

And they're all in their pods with the dials cranked up to 11.

Quick Poll Question

Why haven't aliens visited Earth?

They've figured out happiness / They lack the technology /
No aliens exist / They're already here

Vote at consciousrobots.com/100.

If the author's robotic emotional system could be aware that you have read this far, it would consider this event to have increased the survival chances of the author's genes. And thus, it would be dishing out the dog biscuits to the author's conscious mind.

The author would ask you to leave a review on Amazon.com, but hopefully you're now convinced your actions are as rational and calculated as a robot. So, you're going to do it anyway, to increase the survival chances of *your* genes, aren't you? Luckily, he's made it very easy for you – just scan below, or visit bit.ly/ConsciousRobotsReview.